THE DARTFORD-THURROCK RIVER CROSSING

A PHOTOGRAPHIC HISTORY

PAUL SMITH

First published 2009
Reprinted, 2019
The History Press
97 St George's Place, Cheltenham,
Gloucestershire, GL50 3QB
www.thehistorypress.co.uk

© Paul Smith, 2009

The right of Paul Smith to be identified as the Author
of this work has been asserted in accordance with the
Copyrights, Designs and Patents Act 1988.

All rights reserved. No part of this book may be reprinted
or reproduced or utilised in any form or by any electronic,
mechanical or other means, now known or hereafter invented,
including photocopying and recording, or in any information
storage or retrieval system, without the permission in writing
from the Publishers.
British Library Cataloguing in Publication Data.
A catalogue record for this book is available from the British Library.

ISBN 978 0 7524 4828 2

Typesetting and origination by The History Press
Printed in Great Britain by TJ International Ltd, Padstow, Cornwall.

CONTENTS

	Acknowledgements	4
	Introduction	5
1.	How it all Started	7
2.	The Construction of the First Tunnel	19
3.	The Construction of the Second Tunnel	37
4.	The Toll Plazas and Control Room	53
5.	The Construction of the Bridge	61
6.	Vehicles Used at the Dartford Crossings	89
7.	Methods of Payment of Tolls	101
8.	Badges of Rank	107
9.	Other Photographs of Interest	111

ACKNOWLEDGEMENTS

The majority of the photographs in this book are from the archives of Le Crossing Co. Ltd. Particular thanks go to: Phil Hart of the Department for Transport for giving permission to use the material from the archive, and to the staff of Le Crossing Co. Ltd – Peter Goddin, managing director; Tracey Chappell; Andy Mack, infrastructure manager; Steve Pattrick, structures manager.

The author would like to thank all those who have contributed to this book: Mike Breading, John Burrows, Mark Coomber, Nigel Fishenden, Hilary Fishenden, Carol Fisher, Vic Hicks, Lynne Waterman-Jarvis, Geoff Long, Wilma McLintock, Jim Munn of Munns of Gravesend, Tony Prior, Brian Salter, Kevin Woodger, Colin Genner and the Dartford Museum.

Special thanks go to all other members of staff at Le Crossing Co. Ltd for their help and support.

INTRODUCTION

The original idea of a tunnel under the Thames from Tilbury Fort to Gravesend was put forward in the late eighteenth century. Due to cost the idea was abandoned in 1803. In 1924 the Gravesend Tilbury Crossing was rejected in favour of a crossing between Dartford and Purfleet, and in 1929 the Essex and Kent County Councils promoted a bill for the construction of a tunnel. The construction of a pilot tunnel was authorised in 1936 and completed in 1938. The Second World War prevented progress on the full-bore tunnel but preparations for its construction were revived in 1955. The tunnel was completed and opened in late 1963 at a total cost of £11 million. The toll charged for a motor car at this time was 2s 6d and over the coming years the traffic flow increased to such an extent that a second tunnel was needed. This was built and opened on 16 May 1980. The second tunnel was sited approximately 21m downstream of the first tunnel.

There are three cross passages that link the two tunnels. Emergency standby power is provided by two diesel alternators, each of which will provide enough power in the event of a power cut. Traffic surveillance is achieved by the use of CCTV cameras within and outside the tunnel. Pollution of the tunnel air is monitored by detectors for smoke and carbon monoxide at three positions within the tunnel and an alarm signal is given in the control room of the concentrations, each at a preset danger level. The toll charge for a motor car at this time was 40p.

By the time the second tunnel was open the traffic was coming from further afield with the completion of the M25 orbital motorway. Due to ever-increasing traffic flows (11,000 vehicles a day in 1963, 30,000 vehicles a day in 1980, and by 1985 it had risen to 65,000 vehicles a day), it was felt there was a need for a third river crossing.

Trafalgar House and its partners in a new company called Dartford River Crossing Ltd won the competition to build the third Dartford crossing. Their scheme was to build a four-lane, cable-stay bridge carrying traffic southbound from Essex to Kent. This would then enable the two tunnels to run four lanes of traffic northbound. The work started on building the bridge with the breaking of ground on 2 August 1988. The revenue to finance the building of the bridge came from the collection of tolls to use the existing crossing. For the cementation Cleveland & Dartford Consortium was given the design and construction contract, Trafalgar House Construction Major Projects looked after the roadway's substructures and overall project management and Cleveland Structural Engineers were responsible for the superstructure of the entire crossing. Trafalgar House Technology designed the bridge substructure.

The first major elements of the overall structure were the two caissons which provide the base of the towers for the pylons; these were built in a dry dock in Holland and towed 150 miles across the North Sea to Dartford. The caissons were designed to withstand a 60,000-ton ship at 10 knots and they weighed 110,000 tons. The cables weighed 1,500 tons and were woven at British Ropes in Doncaster. The diameter of the steel cables was 16.5cm and they were the largest made in Britain to date, and were designed to carry the heaviest traffic loading in the world. The steel used to build the bridge was the equivalent to three Eiffel Towers. The length of the bridge was 2,869m and the main span cleared the river by 60m. The third river crossing was named the Queen Elizabeth II Bridge and was opened by Her Majesty on 30 October 1991. The opening of the bridge improved traffic flow enormously.

Following the opening of the bridge, Dartford River Crossing Ltd took further steps to speed up the traffic flow using an electronic toll system called Dart-Tag. Dart-Tag is a device based on a computer chip that is fixed inside the vehicle's windscreen. As the vehicle approaches the toll plaza, the equipment installed at the plaza will read the tag and, provided there are sufficient funds in the account, the barrier will rise providing the motorist with minimum delay. The Dart-Tag system operates in all twenty-seven lanes for a motor car; the toll charge in 1992 was 80p for motor cars.

The toll charge for a motor car from 1 September 1996 was £1. The Ministry for Transport introduced a new charging order from 15 November 2008 and now the toll charge for a motor car is £1.50.

The concession to build the Queen Elizabeth II Bridge and run the crossing by the Dartford River Crossing Ltd ended in 2003 after a successful fifteen-year run. On 1 April 2003 Le Crossing Co. Ltd was awarded a managing contract to run the crossing until spring 2009. (All statistics taken from: Dartford River Crossing Ltd)

Note: *GTG = Grays & Thurrock Gazette*

1
HOW IT ALL STARTED

A brass plaque from the 1920s measuring 8cm x 13cm showing the Kent to Essex ferry-crossing charges. At the top are the two pubs with the ferry boat in the middle. The wording underneath says: 'Original Kent and Essex Ferry, Greenhithe, applications at the White Hart Ferry Office'.

The Sir John Franklin public house at Greenhithe, Kent, formerly the White Hart which used to be the ferry office from the 1820s.

The idea of charging for crossing the Thames at Dartford goes back to medieval times when tolls to use the ferry crossing were managed by nuns. Nuns from Dartford Priory were given the right to operate a ferry service for which they would have charged to raise money for the church. Dartford Museum possesses a toll-charge plaque from the 1820s showing that carriages with two horses were charged as much as 10s to cross, while the cost for lone horses was 1s 6d. Farmers crossing with herds of livestock were charged for each animal, for example 2d each for each sheep and pig.

The ferry crossing at that time was between two pubs, the White Hart in Greenhithe and the Rising Sun in Essex. It was the pubs themselves that ran the crossing as a private enterprise.

The White Hart was on the Kent side of the river, and it is still there but is now called Sir John Franklin. The Rising Sun public house was on the Essex side of the river, but is no longer there.

At a meeting of the Orsett Rural District Council one of the members showed great concern about the fact that the proposed Purfleet to Dartford Tunnel would come out within about 100yds from Stonehouse Barrier, making this a very busy junction (*GTG* 03/01/1925). At a meeting of the West Thurrock Parish Council on 12 February 1926 the chairman, Mr Stapleton, put forward a proposition that they contact representatives of Aveley, Stifford and South Ockenden Parish Councils with a view to forming an urban district council. It would appear that there was some proposal in the air about amalgamating with other councils to form an urban district council similar in size to the present borough council area. During the tunnel

The view from behind the public house at the point where the ferry crossed the river to the Rising Sun public house in Grays, Essex. The pub no longer exists. Note the Queen Elizabeth II Bridge to the left of the photograph (2007).

planning stages, the developers had concerns that the tunnel earmarked for West Thurrock might be poached by Tilbury (*GTG* 13/02/1926).

At a meeting of the Tilbury Urban District Council the clerk reported that he had taken steps for a deputation to go to the Ministry of Transport covering the siting of the lower Thames Tunnel and he was still awaiting a reply (*GTG* 06/03/1926).

At a meeting of Tilbury Urban District Council on 7 February 1927 it was decided by the committee, comprising representatives from Tilbury and Gravesend, to hold a public meeting in March to be addressed by Gravesend's MP, Mr H.W. Lockyer. The chairman remarked that at last Gravesend was 'Waking Up' (*GTG* 11/02/1927).

At a meeting of the Tilbury Urban District Council on 5 March 1927 one of the members of the joint Lower Thames Tunnel Committee urged that they withdraw the joint venture with Gravesend as Gravesend has refused to assist Tilbury in securing the building of the tunnel at that point (*GTG* 12/03/1927). Over this period there seemed to be a bit of fighting over who was going to get it; it would appear that Purfleet did not want to amalgamate with Tilbury Council as there would be a conflict of opinion as to the siting of the tunnel. This was later resolved and the tunnel at Tilbury was shelved although an attempt to get a bridge over that point of the Thames was made in the late 1930s.

On 30 July 1935, at a meeting of the Purfleet Urban District Council, the chairman, Mr E. Hanford, said that the tenders for the Purfleet-Dartford Tunnel had been received by the Ministry of Transport and that preparations are well in hand. Asked if there was a prospect of the tunnel being commenced in the 'reasonably near future', he replied 'yes, I think there is'. A discussion took place about this and Cllr Mr A. Lockyer said jokingly that if the council were to invite the Ministry of Transport to open he might get a move on (GTG 03/08/1935).

On 5 November 1935 at Erith, Kent, Mr Herbert Morrison MP, who later became Home Secretary in the Labour Government, promised that if the Labour Party got a majority in the forthcoming elections, and that Kent and Essex Council Councils played the game, there would be no hesitation in commencing the tunnel. Herbert Morrison denied that London County Council was pressing for a bridge at Woolwich. He further added that if Essex and Kent County Councils had done their duty the tunnel would have been built many years ago (GTG 09/11/1935).

On 10 January 1936 the Ministry of Transport announced that as the result of the conference the previous day, in which representatives of Essex and Kent County Councils and the Ministry of Transport were present, the Purfleet-Dartford Tunnel could go ahead. It was also hoped that work would commence during the current year. Alderman Heather accused the Government of hanging back from starting big schemes as they 'might want all the money they can get from people later on, to pay for the war'. Cries of 'rot' and 'who with' greeted this remark. The discussion was then dropped.

The same paper was discussed at a meeting of the Purfleet Urban District Council and the members were very pleased with the news on 5 March 1936. Purfleet Urban District Council amalgamated with the Grays, Tilbury and Orsett Councils and the business of the local interests in the tunnel moved from the Purfleet to the Thurrock Urban District Council (GTG 18/01/1936).

On 7 April 1936 at the meeting of the Essex County Council the chairman of the Highways Committee reported that the revised scheme for the Purfleet-Dartford Tunnel had been submitted after consultation with representatives from Essex and Kent. A further report from the late Dr D. Haldone stated that the work could be carried under safer conditions with the new revised scheme. Basically the old scheme consisted of a circular tunnel of 10.4m in diameter with a 5.8m-wide carriageway and two 1.4m-wide footways. Under the new scheme it would be 9.1m wide, would be dish shaped and the carriageway would be 6.1m wide but there would be no provisions for pedestrians with the exception of two small footways for tunnel patrols. The new scheme would reduce the cost of the tunnel by approximately £300,000. Cyclists and horse-drawn vehicles would be excluded from using the tunnel.

County Cllr Revd Sorenson MP stated:

> It would be spoiling the ship for a half penny worth of tar ... I have yet to understand that the tunnel is being paid for altogether by motorists. Why try and push the pedestrians off the road or prevent them from passing under the water? ... If you are going to stop young persons and working people who might find the tunnel useful it looks suspiciously like class legislation.

He later said:

> I am standing for oppressed classes of cyclists and pedestrians. If this is a class war against the cyclists I want to know 1) how much will be saved by reducing the tunnel to exclude cyclists and pedestrians and 2) what other provisions will be made for them?

The chairman, County Cllr Webb, said that the reduction was not done to save money but the smaller diameter would make it much safer for the men working under the compressed air conditions. Cllr Webb also stated that the original plans included a toll for pedestrians but he added buses would be provided and the fares would be no greater than the tolls (*GTG* 11/04/1936).

On 13 August 1936 the Minister of Transport, Mr Hore-Belis MP, in conjunction with a committee consisting of representatives from Essex and Kent, announced that arrangements had been made for work to commence on the Pilot Tunnel contract. The contractors Charles Brand & Sons had submitted the lowest of ten tenders.

The specifications of the pilot tunnel were to be 3.6m in diameter and 975m long. There would be two shafts, one on each side of the river, and they would be 5.5m in diameter and each would be about 30.7m deep. The depth of the tunnel would run about 7.6m under the Thames. The Pilot Tunnel and ancillary work would cost about £300,000; the cost of the

Construction of the tunnel-cutting machine, January 1937.

completed tunnel would be approximately £3,200,000. The work on the Essex side of the river would commence at a point near Stonehouse Corner and would cross the Southend & Hornchurch St railway line by means of a bridge (GTG 15/08/1936).

On 19 August 1936 the *Grays & Thurrock Gazette* reported an interview with the consultant engineer Mr David Anderson of the firm of Messrs Mott, Hay & Anderson at their Westminster Offices. D. Anderson stated that he would be seeing contactors during the week with a view to agreeing on a starting date. The first task would be laying access roads, and then sinking the shafts, one on each side of the river, which would be lined with cast-iron sections. The tunnelling would be carried out by a cutter which would be driven through the sub-soil with hydraulic ramps. As it moved forward the sub-soil was driven to the rear and was extracted by men known as miners. These men would be working in compressed-air conditions and the shift would be of short duration. Compressed air was used because it prevented water seepage into the tunnel. Questioned by the *Grays & Thurrock Gazette* the engineer replied that the Pilot Tunnel and ancillary works would need between 200-300 men. They would be divided into roughly two gangs, one on each side of the river. He said that the miners were very skilled men and they would love to be brought in by the contractors, but he added that the remaining 75 per cent would be recruited in the Grays and Dartford areas (GTG 22/08/1936).

Towards the end of September 1936 the tunnel contactors Charles Brand & Sons took over an empty cottage at Stonehouse Corner and converted it into a site office. This led to mostly unemployed men attending the office seeking work, and this resulted in articles in the *Grays & Thurrock Gazette* stating that all jobs would be allocated through the labour exchanges at Grays and Dartford (GTG 26/09/1936).

In March 1937 a report appeared stating that certain ancillary work on both sides of the river had been completed and it hoped that the actual sinking of the tunnel would start at the end of the month. In the same issue of the *Grays & Thurrock Gazette* an article appeared concerning Gravesend Town Council, in which an Alan Date JP wished to consult other councils i.e. Tilbury Urban District, Grays, Urban District and Northfleet Urban District, with a view to forming a committee to approach the Ministry of Transport concerning a bridge access to Thames between Gravesend and Tilbury. Cllr Oaten JP was informed that Tilbury and Grays Urban District Councils had now amalgamated with Purfleet and Orsett Councils forming the Thurrock Urban District Council. The proposal was amended and a committee of five members of the Gravesend Council were elected with a view to confirming with Thurrock Urban District Council. (GTG 13/03/1937).

A report appeared in the *Grays & Thurrock Gazette* on 17 April 1937 when a reporter from the paper interviewed the resident engineer on the site, Mr John F. Hay. He remarked 'The job is now in being'. He further remarked, 'Up to now we have only been making preparations, building roads and getting materials onto the site'. The reporter went on to say that the sight of men working in a hole, although it was not the actual tunnel, was most encouraging and that the most important item on the work schedule that week was the placing of the cutting blade into the shaft. The engineer stated that work on the Essex side was progressing slightly faster than the Kent side. The article also reported that it was hoped the shaft excavation would be in full swing the following week. A very detailed report of how the excavations were being carried out also appeared in the article.

On 27 April 1937 a meeting was held by the Commercial Motor Users Association, chaired by Mr Victor Raikes, MP for south-east Essex. The purpose of the meeting was to protest about the tolls to be charged for using the tunnel. The association had petitioned the House of Lords and had secured some reductions in charges. The maximum rate for a 10-ton lorry would be 6s for a single journey and, as the result of the petition, this had been reduced to 4s. The charge for a 2.5-ton lorry had been reduced from 3s to 2s. Speaker after speaker protested about any tolls on the tunnel (*GTG* 01/05/1937).

In the *Grays & Thurrock Gazette* of 29 May 1937 an article appeared concerning the progress of the sinking of the shafts. The article stated that on the Essex side of the river the shaft would be almost 24.6m deep. The method used was to force a 5.6-m diameter cast-iron, which had a sharp cutting edge, into the soil forcing the surplus soil into the centre of the ring, from where it was cleared to a mechanical shovel. Increased weight on the ring was enough to force it through the soil, but now – at a depth of 12.3m – chalk, a much harder substance, had been found which would require digging out by men using pneumatic picks working under compressed-air conditions.

A short report in the *Grays & Thurrock Gazette* 26 June 1937 stated:

> The huge boiler-like contrivances can now be seen at the top of the shafts. These are the airlocks through which the workmen must pass when they go down and come up from their work. The report also states that the vertical shaft is almost complete and soon our 'eye' will be made in the shaft and then the horizontal excavations can begin.

Building the Dartford Tunnel, January 1938.

On 30 September 1937 a fatal accident occurred at the tunnel, when a Mr Frederick John Williams, aged thirty-two, of Dahlion Cottages, West Thurrock, fell 20m from the top to the bottom of the shaft. He died almost immediately (*GTG* 02/10/1937).

On Wednesday 26 January 1938 the *Grays & Thurrock Gazette* reporter again visited the tunnel and reported that compressed air had been reintroduced into the tunnel after being withdrawn for about a month to enable certain tests to be carried out prior to the final horizontal bore under the river (*GTG* 29/01/1938).

On 3 May 1938 it was reported in the *Grays & Thurrock Gazette* that the 'miners', numbering about fifty, had gone on strike for an increased bonus payment. It was reported that there were six shifts, each of four hours, working under the compressed-air conditions. In the same copy of the newspaper there was the report of an inquest held at Tilbury Police Station on 30 April 1938 regarding the death of Mr Thomas Logan Richards, aged thirty-two, of Athol Road, Dundee. The brief facts were as follows: Mr Richards was employed on the Dartford Tunnel as a minder's labourer. On the night of 11 April 1936, Mr Richards was in the mess room on the Dartford side of the river. A witness gave evidence and said that Mr Richards had missed the boat from Dartford to Purfleet which left about 10.10 p.m. Richards told the witness he was going to make his way towards the jetty to get the next boat to Purfleet. A lock-keeper working in the yard stated that at about 10.15 p.m. he had heard shouting and saw the deceased struggling in the water. He threw a lifebuoy but the struggling man could not grasp it and he was swept away (*GTG* 07/05/1938).

On 16 August 1938 a *Grays & Thurrock Gazette* reporter visited the site, although he could not go into the tunnel because of work being carried out in compressed air, but he was given some statistics. The miners were now 415m from the bottom of the shaft and had crossed the county boundary by 15m. They were in the deepest part of the tunnel which at that point had a gradient of 1 in 28, and they were 26m below the high-water mark. On the Kent side they had come across several difficulties and their progress was not as good. It was said that if things carried on as they were, the two gangs would meet up in mid-October (*GTG* 20/08/1938).

At noon on 4 October 1938 the tunnels started from each bank were 'junctioned', meaning that they met up in exactly the right position. It began on Thursday week, on which day the drilling was stopped on the Kent side and what is known as a 'box' was driven for a length of 6m. The box was 1.8m high and 1.2m wide as against the tunnel's diameter of 3.6m. The drilling from the Essex side continued until Saturday when they were only about 6m apart. Then a hole 0.6m wide was bored from Essex to the Kent side. This is normal engineering practice as it equalises the pressure in both tunnels. The tunnelling was then continued from the Essex side, resulting in the complete breakthrough at noon on 4 October 1938 (*GTG* 08/10/1938).

On 29 November 1938 the Minister of Transport, the Rt Hon. E. Leslie Burgin MP LLD, and a number of other distinguished gentlemen inspected the Pilot Tunnel.

The first crossing of the River Thames at Dartford was a twin-lane bored tunnel which opened in 1963 as a local link between the A2 and A13. A joint committee from Kent and Essex County Councils was responsible for its operation and maintenance. Construction took five years as work was hampered by difficult tunnelling conditions in the chalk. On opening, traffic was over 4 million vehicles per year or 12,000 vehicles per day. This increased steadily over the following years until in the early 1970s it was running at over 10 million vehicles per year. The joint committee decided that traffic levels warranted a second tunnel of similar capacity and

Artist's impression of the tunnel from Dartford to Purfleet printed in the *Modern Wonder Magazine* on 26 March 1938. The following inscription printed with the drawing did not become a reality:

> Sectional view of the new tunnel being built under the Thames from Dartford (left) to Purfleet at a cost of about £3,500,000. It will be ready for use in 1941, and it is hoped that the boring of the pilot tunnel will be finished by next September. The main vehicular tunnel will have a diameter of 9.2 metres, and the distance from bank to bank at this point is 871 metres. Insets are seen 1. The circular shield used for boring which has a diameter of 9.5 metres and a length of 4.8 metres. It includes sliding platforms to excavate before the cutting edge, an erector to hoist segments into position, trams to remove excavated material, and peripheral rams to force cutting edge into ground. 2. The finished tunnel. 3. The vertical shaft leading to working shafts.

construction began in 1972. Again, the difficult ground conditions prolonged construction and the second tunnel did not open until 1980.

Traffic using both tunnels now totalled 11 million vehicles per year or 30,000 vehicles per day. Capacity was not a problem until 1984 when sections of the M25 London Orbital Motorway began to be opened. In 1984 the county councils asked the Government to take over responsibility for the tunnels as part of the national motorway network. The Secretary of State for Transport commissioned a traffic report, which forecast that the existing tunnels would be overloaded to an unacceptable degree by the early 1990s.

In March 1986 the Department of Transport gave permission to Dartford River Crossing Ltd to take over the two existing tunnels, discharging the outstanding debts of the Kent and Essex County Councils, and to design and build a new four-lane cable-stayed bridge and to maintain and operate the combined crossing for a maximum period of twenty years.

Work on the bridge commenced in August 1988 after the company took over the operation of the existing tunnels under the terms of the concession agreement, and it was officially

Sir Frank Whittle had the inspiration in the late 1920s to replace piston-driven propellers with a gas-turbine arrangement. This was to become a jet engine. In 1936 Sir Frank founded the company Power Jets to develop his ideas. The testing of the gas-turbine engines was done on the River Thames at Dartford in the 1940s where the tunnels are now.

opened by Her Majesty the Queen in October 1991, on time and within budget. The Queen Elizabeth II Bridge marked the first fully privatised infrastructure project in the UK in the twentieth century. Since the opening traffic has grown by some 75 per cent and traffic levels in 2006 were in excess of 50 million vehicles per year.

The Traffic Control Operations Centre, overlooking the toll plazas on the Kent side, monitors and supervises a sophisticated array of equipment ensuring the safe and efficient flow of traffic through the crossing. CCTV cameras provide an instant and continuous view of traffic conditions on both approach roads, in the tunnels and on the bridge. A radio system allows communication with staff and an internal telephone system allows communication throughout the crossing and includes emergency phones at 50m intervals in the tunnels and on the bridge.

This communication network enables a rapid response to any incident and ensures maximum standards of safety are maintained at all times.

The tunnel's monitoring systems measure vehicle exhaust emissions to ensure the correct level of ventilation is maintained. On the bridge, sensors buried within the road desk measure surface and air temperatures; anemometers are mid-span and on the pylons, which measure wind speed and direction. In high winds the system allows a progressive series of measures to be applied to restrict the speed and position of vehicles on the bridge, and at very high wind speeds to close the bridge and send all the traffic through one of the tunnels. Within the tunnels, lights are fed alternatively from the Kent and Essex electricity supplies in order that if either supply fails, the tunnels will still remain evenly lit along their lengths. Lighting on the bridge had to be designed with aircraft and shipping in mind as well as road traffic. Street lighting illuminates the road deck at night; in addition aircraft warning lights are located on the pylons and navigation lights and radar reflectors mark the shipping navigation channel. Ventilation of the tunnels is critical for users both during peak-traffic times and in an emergency situation, such as a fire within a tunnel. Fresh air is generated by fans situated in ventilation buildings at ground level on both sides of the river and fed into the traffic space by means of under-road ducts which extend between the two ventilation shafts. Exhaust fans are also situated in the same buildings, above the crown of the tunnel, to take away foul air. If a fire did take place, jet fans located in the tunnel roof provide a high forward-air flow to prevent smoke and fumes affecting users behind the incident. Pumping systems ensure that any water seeping into the tunnels and rainwater running off the bridge, is diverted after treatment into the River Thames.

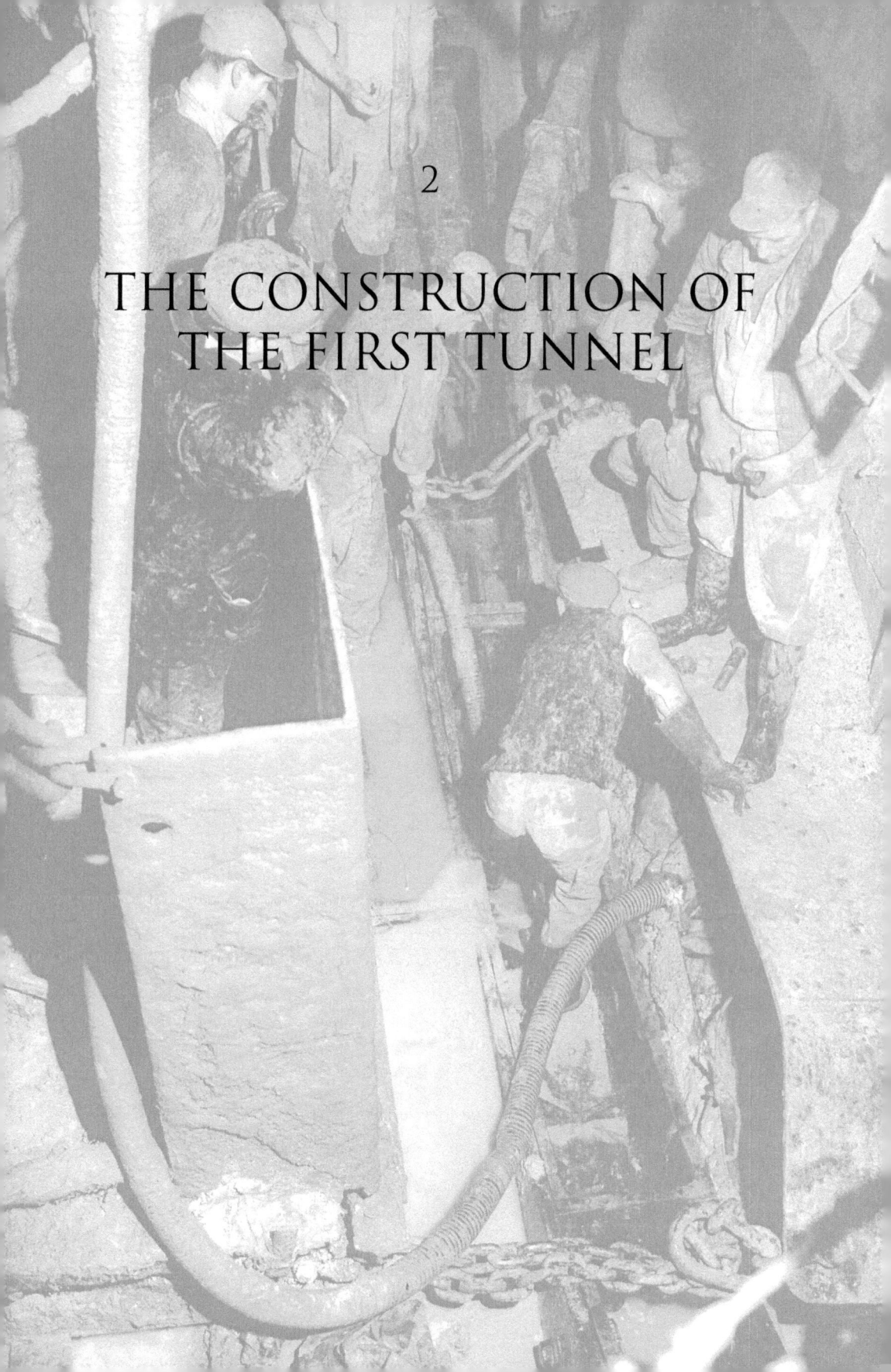

2
THE CONSTRUCTION OF THE FIRST TUNNEL

View across the River Thames at the location where the first tunnel was constructed.

Construction work being carried out on the Dartford Tunnel from the mid-1950s.

Construction work being carried out on the Dartford Tunnel from the mid-1950s.

Construction work being carried out on the Dartford Tunnel from the mid-1950s showing forty cutting-face pusher jacks weighing 125 tons of thrust each.

Above and below: Construction work being carried out on the Dartford Tunnel from the mid-1950s.

Above and below: Construction work being carried out on the Dartford Tunnel from the mid-1950s.

Construction of the tunnel lining at its entrance from the mid-1950s.

Construction workers preparing the ground for the first tunnel.

Above and below: The tunnel and approach roads at various stages of construction.

Above and below: The tunnel at various stages of construction.

Control room and offices on the Kent side.

An officer standing at the entrance to the building, 1963/4.

First toll booths and uniformed toll-booth operators, 1963/4.

Toll-booth operator serving a motorist, 1963/4.

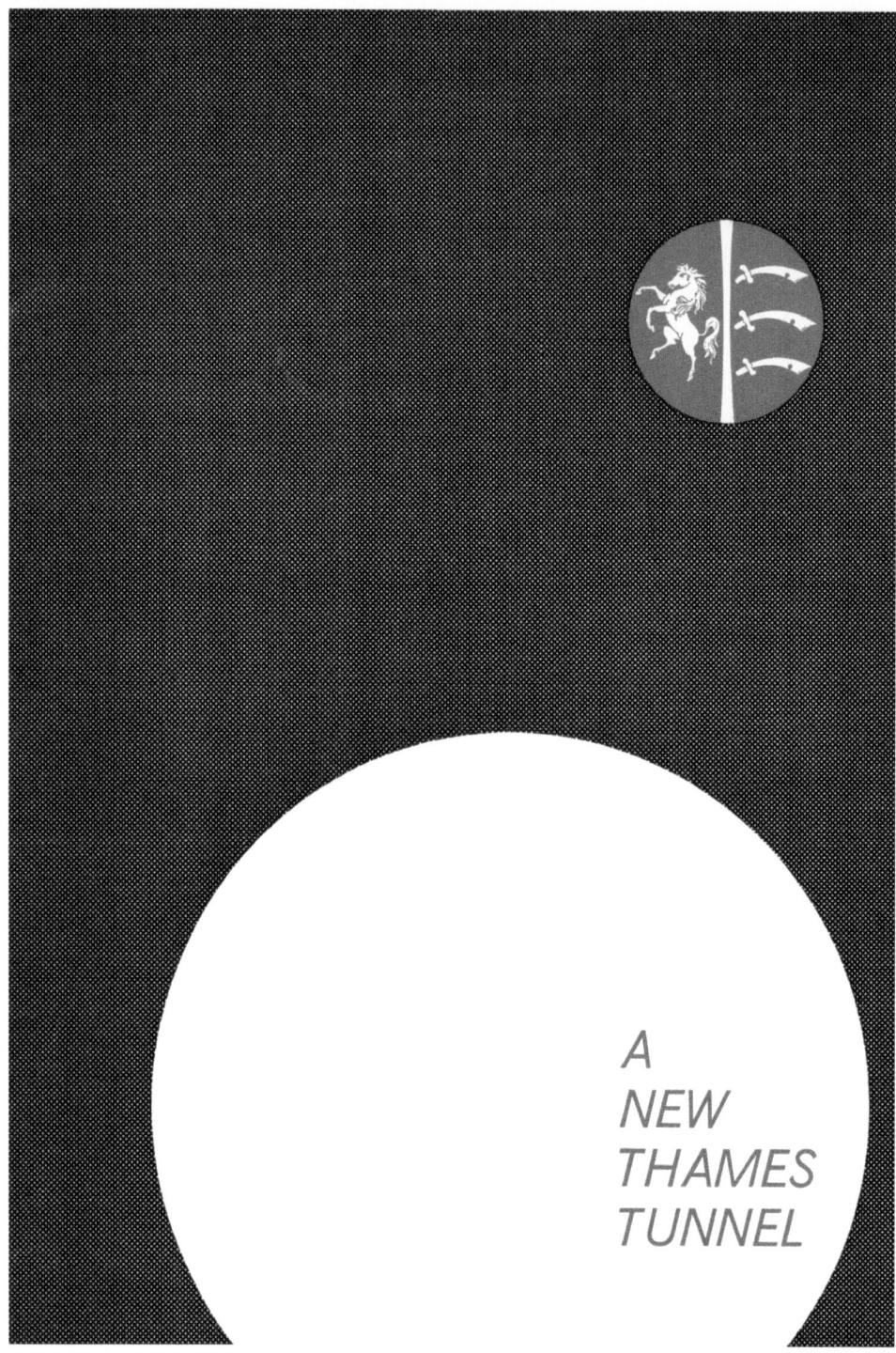

This booklet was produced for the opening of the first tunnel in November 1963. (From Dartford Tunnel Joint Committee – DTJC).

View from the roof of the control room looking north towards the tunnel. Toll plaza in foreground, 1963/4.

Facts about the tunnel:

Length (between portals)	1,432m
Length (including approach roads)	3¾ miles
Total cost of project	£11m
Diameter of tunnel lining:	
cast iron	34,000 tons
metallic lead caulking	160 tons
number of flange bolts	260,000
Materials used in pre-treatment of the gound:	
from the pilot tunnel, Portland cement	10,000 tons
from surface, clay cement and clay-chemical grouts	2m galls
Width of tunnel carriageway	6.4m
Minimum headroom in tunnel	4.9m
Maximum road gradient in tunnel	1 in 28
Lowest point below high-water level	30.7m
Fresh air supplied hourly through ventilation system (peak hours)	1,018,80cu.m

Essex Point looking north, 1963/4.

Vehicles approaching the tunnel from the Essex side, 1963/4.

Entrance to the newly completed tunnel – note the honeycomb section of roofing.

Left and below: First books, issued 1963.

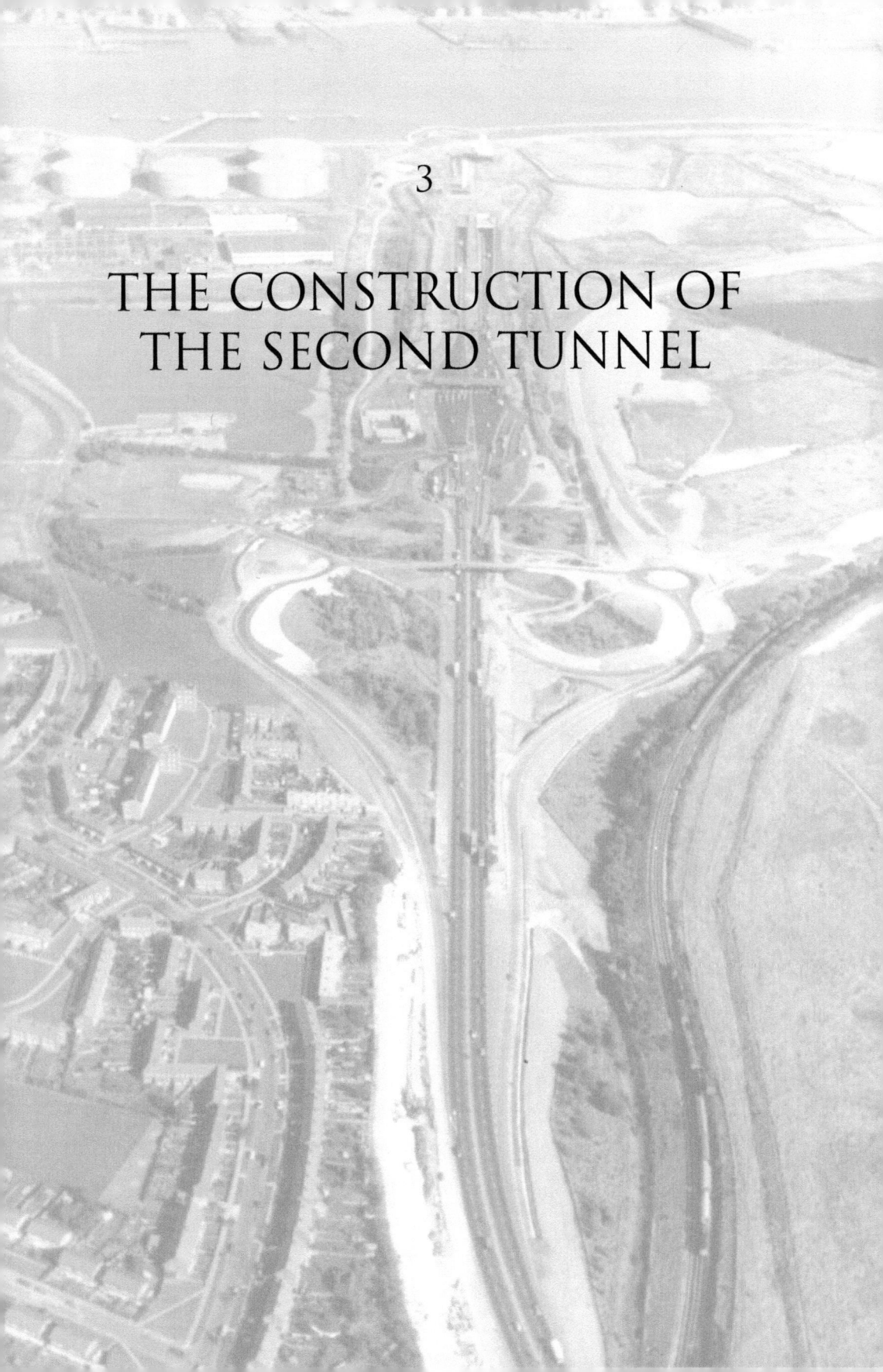

3
THE CONSTRUCTION OF THE SECOND TUNNEL

The breaking of earth on the Essex side to begin construction of the second tunnel in the mid-1970s.

The Kent site – a general view of the toll-plaza area looking south, 1975.

The Kent site – an internal view of the compressor house used in the construction of the second tunnel, looking south.

The Kent site – engine control panel in the multi-motor-panel DA room, Kent West vent.

The Kent site – an external view of medical lock No. 1, 1975.

The Kent site – an internal view of the doctor's surgery and equipment, 1975.

The Kent site – a general view of escol panels in the cut and cover section, looking north, 1979.

The Kent site – a general view of the painting to the composite lining in the main tunnel, looking south, 1975.

The second Dartford road tunnel beneath the River Thames marked an important stage in the development of the road system in the south-east of England. Its function was to improve traffic facilities between Kent and Essex by increasing the present capacity of the existing crossing by allowing one-way or tidal traffic flow through either tunnel.

The crossing forms a direct link between the A13 (London to Southend) and A2 (London to Dover) roads and provides continuity of the M25 motorway north and south of the Thames, about 17 miles to the east of London. The crossing is available for the transit of all classes of vehicles including bicycles, but is not open to pedestrians. High explosive and a few other especially dangerous goods are not allowed passage while other dangerous goods are subject to escort or restriction to certain weights and/or quantities.

The Dartford Tunnel Joint Committee, appointed by the Essex and Kent County Councils under an Act of Parliament, had the responsibility of operating and maintaining the tunnels, approach roads and associated buildings. It was estimated that the tunnel's capacity of 31 million vehicles per year would be achieved by the end of the century. The committee stated that the tolls, after deduction of operating expenses, would be devoted to the payment of interest and the reduction of the capital debt associated with the construction of the crossings. The toll at that time for a private car was 35p for the single journey.

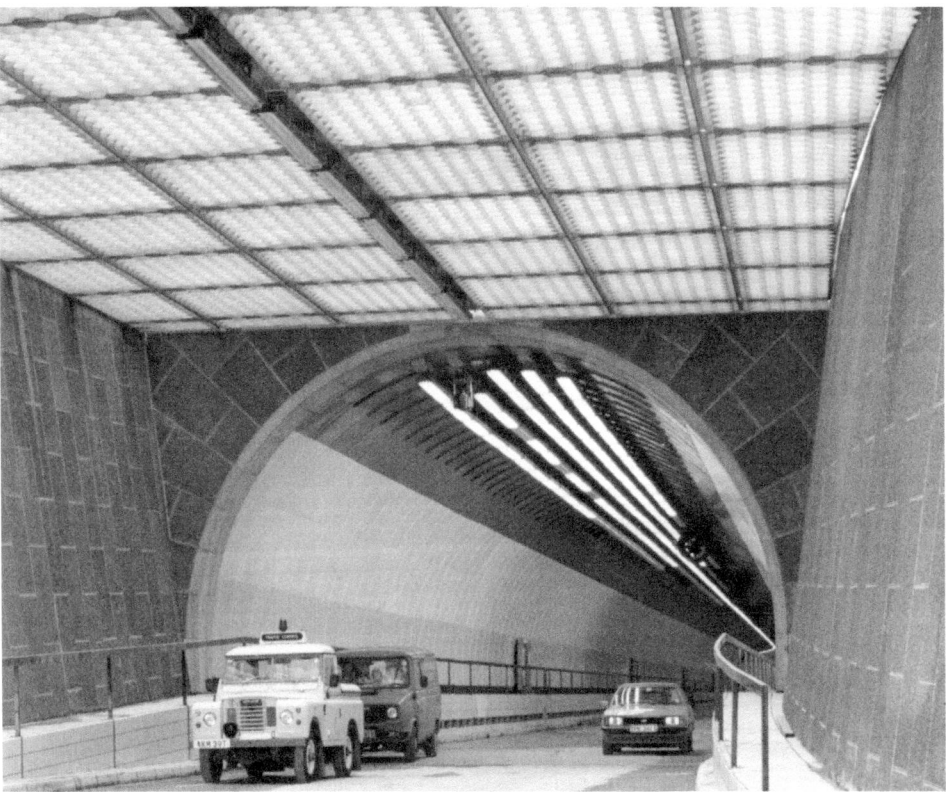

The Kent site – a general view of the second tunnel.

The east-tunnel approach on the Essex side, 1980.

The Essex side facing south towards the east tunnel, 1980.

The Essex rover track leading to the vent building.

A picture of the Kent site taken from the flood bank showing the south side of the ventilation building, 1977.

The Kent site – a diesel generator in the vent building, 1978.

Extractor fans in ventilation buildings.

A ventilation shaft, taken from inside the tunnel roof.

One of many jet fans mounted at each end of the tunnel.

An aerial photograph of the tunnel openings looking towards Essex.

An aerial photograph of the crossing looking from Essex to Kent

An aerial photograph of the tunnel approach and exit roads. Note the Temple Hill housing estate on the left of the photograph, 1980.

4

THE TOLL PLAZAS AND CONTROL ROOM

North plaza toll booths.

Construction of south plaza toll booths.

Construction of the roadway leading to the tunnel looking south. Note the hazardous goods-vehicles' marshalling area to the right.

Construction of the roadway and south plaza on the Kent side looking north. Note the marshalling area to the left.

One of the general manager's yearly reports.

South plaza auto tolls at night heading into Kent.

Automatic toll lanes on the newly built south plaza heading into Kent.

Aerial shot of both toll plazas looking south. Note the control room on the right of the photograph.

Control-room rear, pre-1990.

An early control room.

The controller in a revamped control room, Kent side, 1991.

5
THE CONSTRUCTION OF THE BRIDGE

The logo of one of the companies who built the Queen Elizabeth II Bridge.

Trafalgar House Public Ltd Co., and its partners in a new company called Dartford River Crossing Ltd, which commenced work in August 1988 on a new 2,700m cable-stay bridge was to be Europe's longest of its kind.

Under the terms of a Concession Agreement awarded by the Secretary of State for Transport, Dartford River Crossing Ltd took over the management and operation of the Dartford Tunnel to construct the new bridge and operate all three crossings for a period not exceeding twenty years. The company then handed back the crossings to the Department for Transport, free of any debt.

The new bridge was built by Trafalgar House Construction Companies – Cementation Construction and Cleveland Bridge & Engineering. The four-lane bridge, which carries southbound traffic across the Thames between Thurrock and Dartford, was due for completion in February 1991. The third Dartford Crossing was the first major transport infrastructure project in Britain to be wholly financed by the private sector. When opened in 1991, the new crossing doubled the present capacity and greatly reduced the congestion at this point of the M25 orbital motorway.

An artist's impression of the Dartford River Crossing. (The Trafalgar House Group)

View of model bridge in the wind tunnel.

Peter Bottomly, Parliamentary Under-Secretary for State, left, and Lord Rockley, Chairman of Dartford River Crossing Limited, sign the concession agreement for the operation of the Dartford Crossings. Mr Bottomly's symbolic 'first toll' of 60p can be seen, centre, on the base of the Dartford Bridge model, 1987. (The Trafalgar House Group)

View of bridge caisson being towed by tugs across the North Sea.

Aerial photograph of the bridge construction on the Kent side. The control room is on the bottom left of the photograph.

Aerial photograph of the bridge construction on the Kent side showing all the road-deck supports in place. Note the two ventilation buildings for both tunnels on the left of the photograph.

Aerial shot of the bridge construction on the Kent river bank.

Aerial shot of the bridge construction on both sides of the river looking from Kent to Essex.

Close-up of the bridge construction on the south side, looking from Kent to Essex.

General view of north viaduct on the Essex side, looking south.

The bridge construction Essex side, looking north. Essex Control Point in the foreground with petrol tankers waiting for escort through the tunnel.

The two ends of the road deck meeting in the middle awaiting the final section, 1991.

The laying of the top surface to the road deck.

Transport Minister Malcolm Rifkin tightens up the centre bolts on the new bridge crossing the Thames, 1991. (Photograph by kind permission from *The Daily Mirror*)

Greasing the cables, July 1991.

The construction of the emergency escape lane situated on the Kent side looking north, August 1991.

The construction of the emergency escape lane situated on the Kent side looking south, August 1991.

The completion of the Queen Elizabeth II Bridge.

The blessing of the bridge prior to opening on 6 September 1991.

The Queen Elizabeth II Bridge at dawn.

The Royal Opening of

The New Bridge

Programme of Events
Wednesday 30th October 1991

The front cover of the 'Programme of Events'.

LUNCHEON
TO COMMEMORATE THE
ROYAL OPENING
OF
THE QUEEN ELIZABETH II BRIDGE

WEDNESDAY 30TH OCTOBER 1991

The front cover of *The Luncheon*.

A commemorative cover showing Her Majesty the Queen opening the bridge.

The opening of the Queen Elizabeth II Bridge by Her Majesty the Queen.

The first tunnel opened on 18 November 1963:

Length	1,446m
Diameter of tunnel lining	8.6m
Width of tunnel carriageway	6.4m
Minimum headroom in tunnel	5m
Maximum road gradient in tunnel	1 in 28
Lowest point below high-water level	30.7m
Fresh air supplied hourly through ventilation system (peak hours)	1,018,800cu. M
Total cost of project	£13m

The opening of the Queen Elizabeth II Bridge by Her Majesty the Queen.

Monument at the base of the bridge, Essex side, 1991.

The second tunnel opened on 16 May 1980:
This tunnel is sited approximately 21m downstream of the first tunnel:

Length	1,435m
Width	10m
Carriageway width	7.3m
Height	5.03m
Cost of project	£45m

The third crossing (The Queen Elizabeth II Bridge) opened on 30 October 1991.

The project started in August 1988 and was undertaken by a consortium of Cementation Construction and Cleveland Bridge Companies, both wholly owned by Trafalgar House. It is the second largest cable-stayed bridge in the world, having a main span of 450m and a fully loaded navigation clearance of 62.5m above the River Thames. The total length of the main span and the two approach viaducts is 1.84 miles. (The largest bridge is in Vancouver, Canada).

There are 112 cables. The longest is 225m with a diameter of 164mm and weight of 33 tons. The cables are coated with special aluminium base grease called Metalcote A to protect them from rusting. The cables are formed from spinning together hundreds of smaller individually galvanised cables. The pylons are 136m high and there is a 'two-man' lift in each. On a clear day it is possible to see six power stations from the top of the pylons.

Some 800,000 high-strength friction grip bolts have been used on the structure. The main foundations in the river, which support the pylons, are two concrete caissons. These were built in a dry dock in Rotterdam and floated over the North Sea. They can withstand an impact of a ship equivalent to the *QE2* hitting them square on whilst travelling at a speed of 10 knots.

All of the steel girders on the completed structure were painted with six coats of paint, which should last for twelve years before it needs repainting. The cost of the project was approximately £86 million.

Other Miscellaneous Facts:

Total number of lanes	Bridge: 4 (taking all southbound traffic)
Total number of lanes	Tunnels: 4 (taking all northbound traffic)
Maximum road speed	50mph for bridge and tunnels
Number of toll booth	14 northbound
	13 southbound
Total traffic capacity for the whole crossing	135,000 vehicles
Design life of the bridge	120 years
Type of bridge	Cable-stayed

(Information from the Dartford River Crossing Ltd)

Aviation warning light mounted on top of the steel pylon.

Close-up of the aviation warning light.

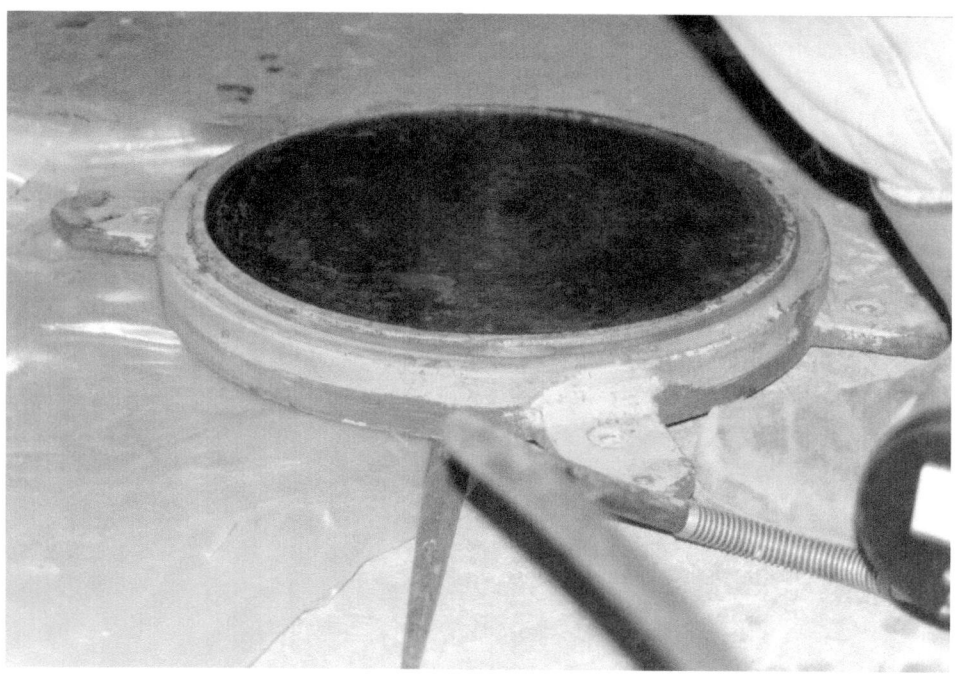

Above and below: One of the many bearings on the bridge which allow the road deck to move in high winds or heavy traffic. These bearings are placed underneath the road-deck's iron work and the bridge supports.

One of the cable anchorage points mounted on the side of the road deck.

Anemometer on top of one of the steel pylons measuring wind speed.

Looking down at the main span from the top of south pylon. Essex is in the background.

Looking south into Kent from the top of the south tower – note the entrance to the tunnels in the top right of the photograph, 1991.

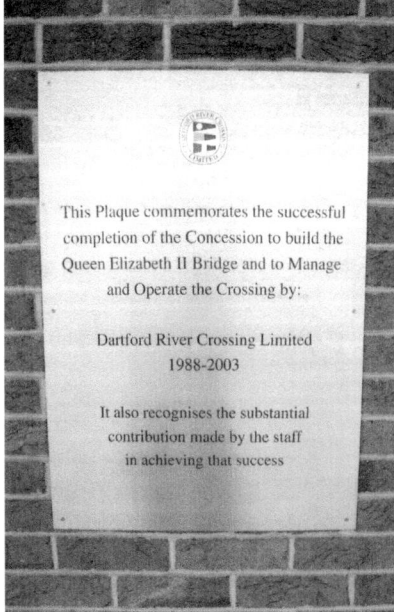

The plaques to commemorate the openings of the three river crossings.

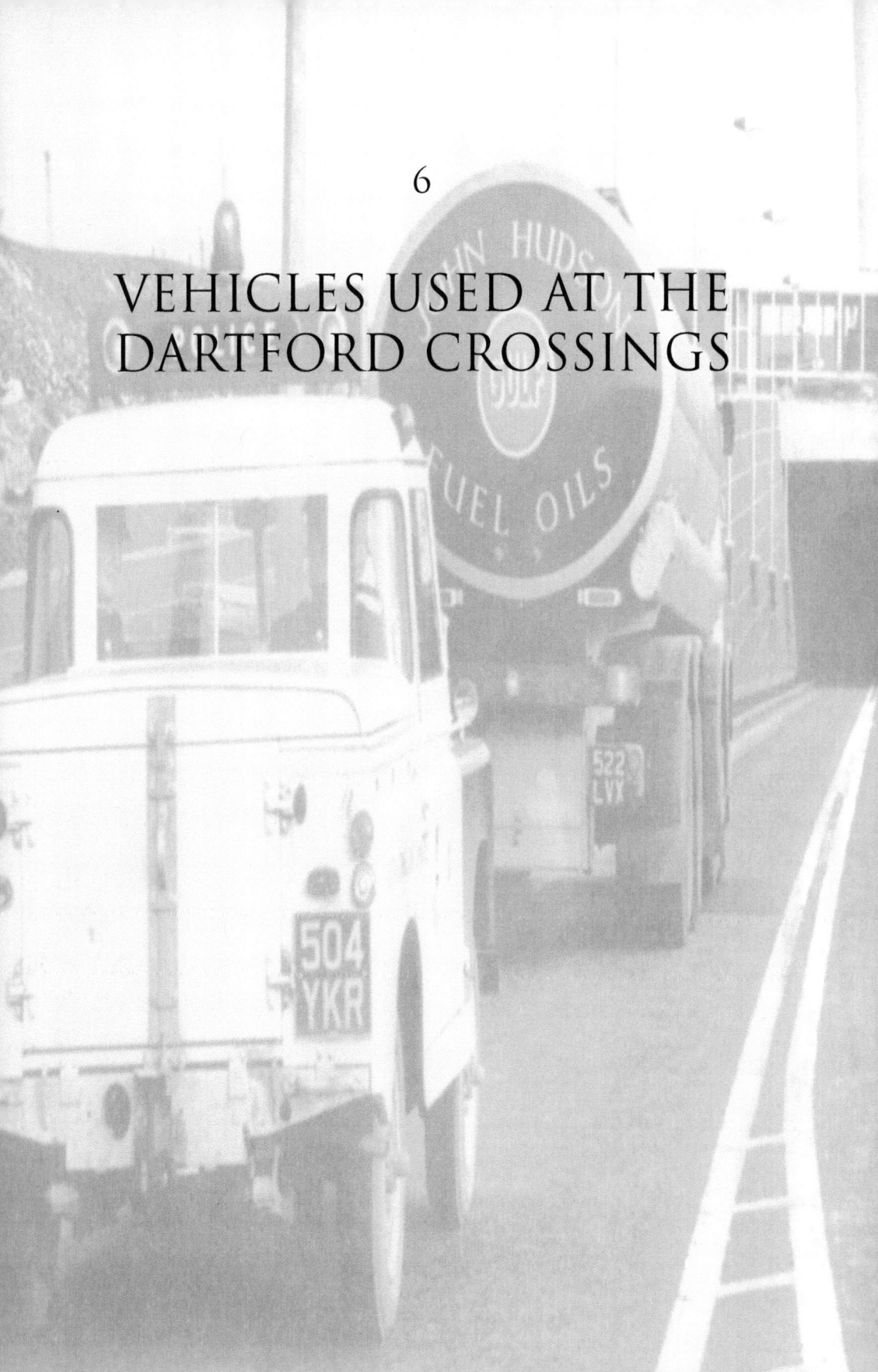

6
VEHICLES USED AT THE DARTFORD CROSSINGS

These vehicles were painted in the old colours of cream for the body and green for the crane and fenders. This colour scheme had to be changed to yellow for all the breakdown and maintenance vehicles by order of the council. This vehicle and an identical one were owned by the Dartford River Crossing Ltd up until 1992/93, when they were sold off. Both scammels are being restored by their new owners.

The heavy-duty Euclid articulated towing vehicle was one of three Euclids owned by the Dartford Tunnel Joint Committee. Two were in use in the 1980s but the third, which was in white, was in poor condition and was probably used for spare parts.

For a while, recovery vehicles carried no registration numbers or tax discs.

Recovery vehicles later ran on trade plates before receiving Q plates, as shown on this vehicle in 1988.

Heavy recovery vehicle pre-1988. Note the Dartford Tunnel Joint Committee logo.

Later Scammell heavy recovery vehicle – note the bracket on the front for the fitting of a snow plough or brush attachment. This photograph, taken in 1995, shows the vehicle with the Dartford River Crossing Ltd logo.

Foden heavy recovery vehicle (DT19). Note all breakdown and maintenance vehicles' colours were changed from yellow to white, 1995.

Iveco low-loader light recovery (DT20), 1995.

Iveco bucket (DT11), 1995.

Iveco scissor lift (DT13), 1995.

Mercedes Unimog (DT14) used for de-icing using special de-icer which does not damage the road surface. It was also used as a snow plough and crane. This is a 1992 vehicle.

Bicycles by bus! One of five specially designed cycle-carrying double-deckers for operation by London Transport on behalf of Dartford Tunnel Joint Committee through the new mile-long Dartford-Purfleet Tunnel, from which cyclists are normally banned. Appropriately the vehicles are based on the Thames 'Trader' PSV chassis, which has been modified slightly to take the special 30ft × 8ft Strachans bodywork. The main open lower-deck compartment accommodates twenty-three bicycles; tricycles and tandems etc go in the rear compartment and there are seats for thirty-three passengers in the upper saloon. Will these ingenious specialists remain unique or will the idea be adapted for other purposes – such as for routes replacing withdrawn railway services where increased luggage accommodation is needed? (Ford Motor Co.)

Ticket machine used on the Dartford Tunnel cycle bus. The price of a single ticket in 1963/4 was 6d.

The first Land Rovers were used by the crossing for escorting dangerous goods. Wide loads were marked 'Police' – the traffic staff in those days had special constable status. These vehicles were painted cream with dark-green front wings.

Land Rover from the mid-eighties with Dartford Tunnel Joint Committee logo, with the then general manager's name (R.L. Jones) on the sills.

Land Rover from 1997, with Dartford River Crossing Ltd logo, was used for site-security purposes. Note the spotlight mounted on the roof.

This patrol Land Rover parked in the yard, showing the Le Crossing Co. Ltd logo covering the old Dartford River Crossing Ltd logo on the doors. Le Crossing Co. Ltd took over the managing of the crossing for the Highways Agency on 1 April 2003.

Dartford River Crossing Ltd patrol Land Rover showing all the equipment that they carry, 1991.

Land Rover Freelander in the new Highways Agency livery.

The latest patrol Ford Rangers used at the crossing for escort duties, 2007.

Rack fitted to the rear of the crossing patrol vehicles to convey cycles through the tunnel or over the bridge.

7
METHODS OF PAYMENT OF TOLLS

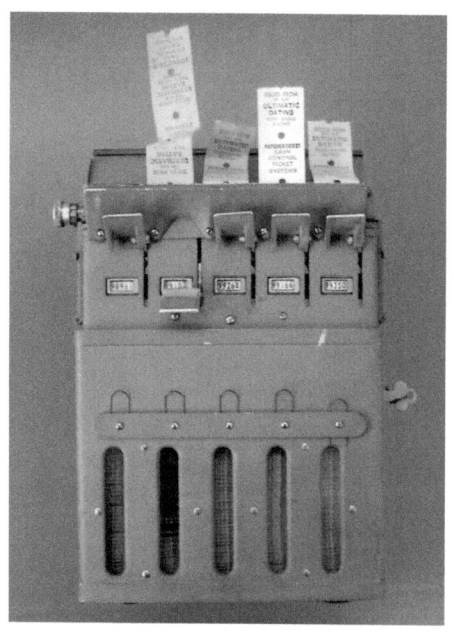 

Ultimatic ticket machine. These machines were used by toll-booth operators in the toll booths up until 1 September 1988, when the Dartford Tunnel Joint Committee became Dartford River Crossing Ltd. They were used in the event of a power cut when an account of toll collecting could not be done electronically. This is a manual machine which issued numbered tickets to motorists for their tolls. On issue to the toll operators by toll clerk, he would log the ticket numbers of the first and last tickets issued to keep account of the number of vehicles through and the money taken. (Dartford Museum).

A pre-paid ticket used by motorists for the payment of tolls up until the early 1990s.

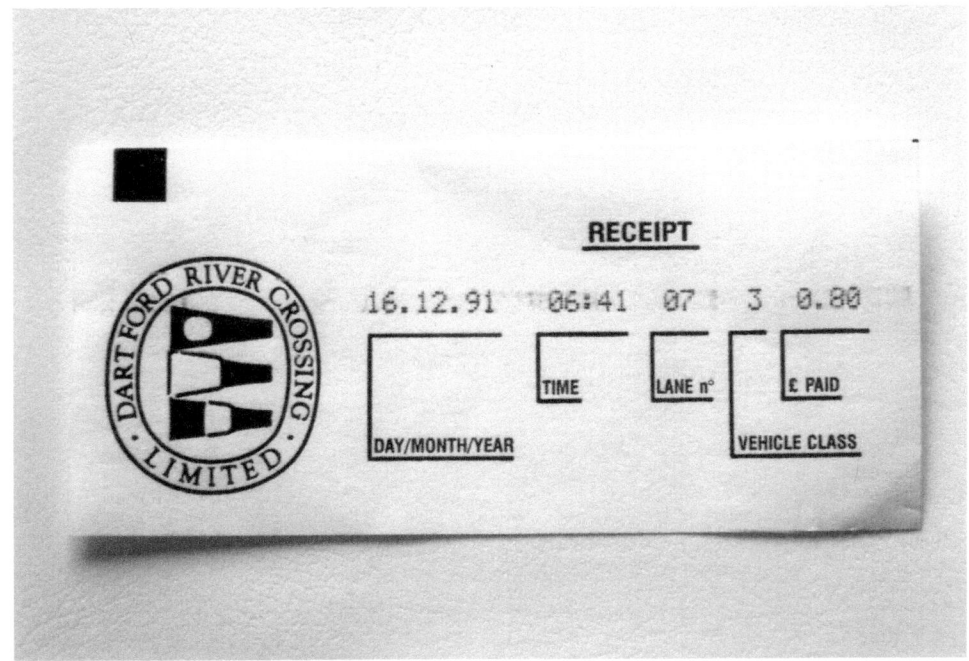

A printed receipt for a car toll of 80p, dated 16 December 1991. The giving of receipts for tolls was trialled for a few weeks but was considered impractical.

Following the opening of the Queen Elizabeth II Bridge in October 1991, Dartford River Crossing Ltd took further steps to ease the journey across the Thames, with a £2 million investment in Europe's most advanced electronic toll system. Dart-Tag is a computer-chip-based device similar in size to a tax disc and fixes to the inside of the windscreen. As the vehicle slows down on approach to the toll booths, equipment installed at the toll plaza will automatically scan the personalised Dart-Tag inside the windscreen. Providing the holder's account is in credit, the barrier will start to rise automatically, allowing the vehicle to pass through almost without stopping. Car drivers using Dart-Tag also benefit from the exclusive use of one lane in either direction, as the far right-hand lane on both toll plazas is reserved for car drivers only using Dart-Tag. (Information courtesy of the Dartford River Crossing Ltd)

View northbound of dedicated Dart-Tag lane for cars only.

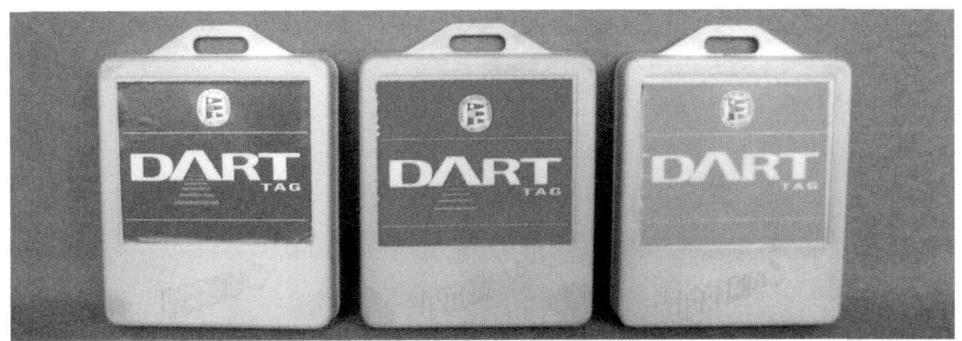

First type of Dart-Tags used which can appear in three different colours – blue for a car, red for a van and yellow for an HGV. These tags were used in conjunction with the readers, shown here mounted on poles at the side of each lane.

A few years later the style of the Dart-Tag changed to the type shown here, and the readers were moved from the side of the lanes to being mounted on the underside of the canopy in the centre of each lane.

Automatic toll machine for cars only.

The leaflet showing the change to toll charges in 2003.

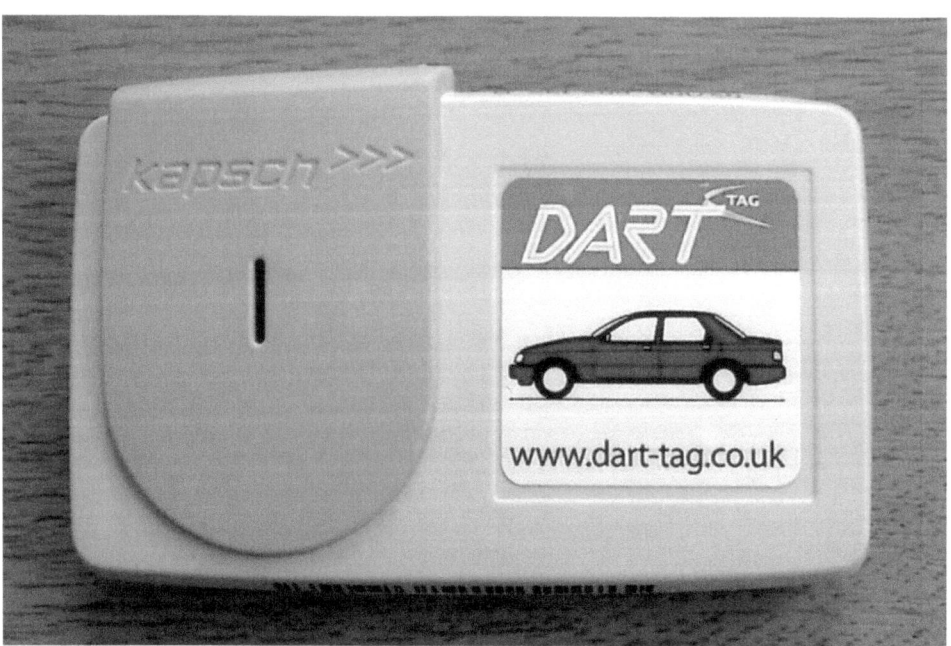

This is the latest style of Dart-Tag used from 2007.

8
BADGES OF RANK

The full range of breast badges worn on the uniforms of the security officer, toll-booth operator, traffic officer, senior traffic officer and traffic-controller staff from the 1960s to the present-day. (Le Crossing Co. Ltd)

The cap badges.

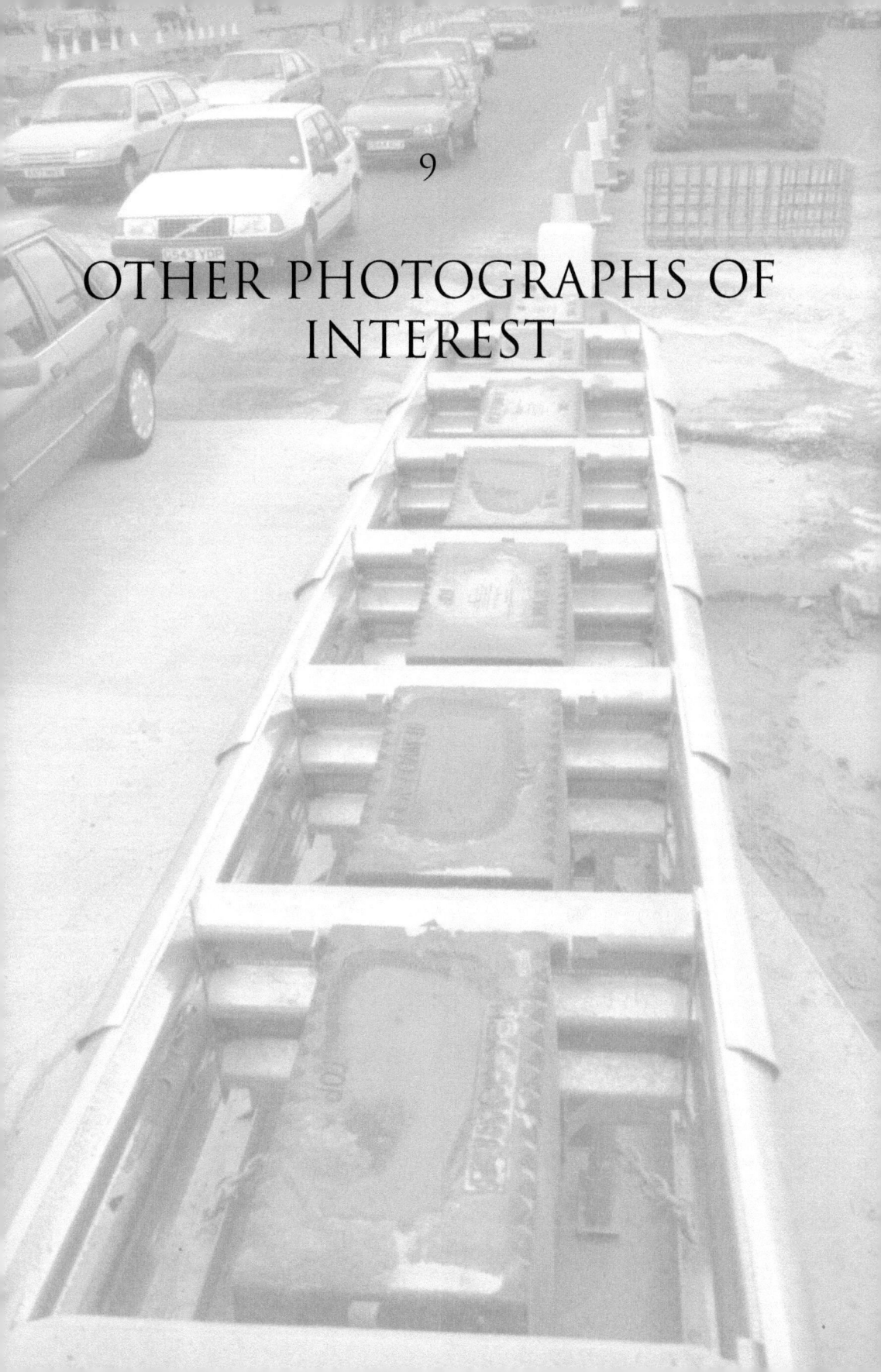

9
OTHER PHOTOGRAPHS OF INTEREST

Left and below: Construction of shock-absorbing crash-protection barriers at the front of each lane protecting the toll booths.

A line-up of the six London buses adapted and used in 2002 for use in both tunnels to replace and upgrade the strip lights that run along the full length of both tunnels.

Air monitors, one fitted at each end of both tunnels.

Dartford-tunnel patrol Land Rover holding traffic at the exit of the Tanker Bay (Kent side) allowing up to six loads to be escorted through the tunnel.

Dangerous traffic book used by traffic officers – this is a list of all United Nations numbers. These numbers determine whether a dangerous load needs escorting through the tunnel or is banned from using the tunnel.

Tankers at night about to be escorted through the tunnel.

Metal height danglers that used to hang from a gantry outside the control building on the Essex-side entrance to the southbound tunnel. These were raised or lowered using a handle to the predetermined height. If over-height lorries hit them the lorries had to find another route.

The result of two cars trying to enter the same lane together – the other car has already been removed.

A crane that has tipped over during the construction of a bridge support.

Outside contractors (Industrial Water Jetting Systems) cleaning the tunnel during the night when one or other of the tunnels is closed.

Pre-cast deck units on casting bed. In the 1990s it was decided that both the tunnels needed their road decks replacing. This was going to be a major operation so needed a lot of prior planning. The road deck was replaced section by section and these pre-cast sections were constructed on site. A section of old road deck was cut and brought out of the tunnel and a new section was taken in and fitted. In 1997 the east tunnel's road deck, which was 1,000m in length, was replaced, and in 2000/01 the west tunnel's road deck, which was 1,400m in length, was replaced. The cost of the entire operation was about £20 million.

This photograph shows the specially made cradle that was used to insert the sections of new road deck.

Crossing Control Point in the marshalling area on the Kent side, 2007.

View of the marshalling area on the Kent side, looking north, shows the old control point on the right of the photograph and the new one on the left. This area is where all dangerous loads and wide, long or high loads pull in for checking before being allowed to go through the tunnel.

The bridge which acts as part of the route for the new Fastrack bus service to the Bluewater Shopping Centre in Kent, November 2007. It goes over the entrance to the two tunnels and under the bridge.

The railway line at Thurrock going under the Queen Elizabeth II Bridge towards Grays in Essex, 2007. This line takes the Eurostar train from St Pancras Station in London to Paris and Brussels.

The new Dart-Tag office opened in 2008 to cater for Dart-Tag customers.

The information point on the approach road to the offices.

A toll-collector's view looking south.

A toll-collector's view looking north.

A stick of rock showing a picture of the Queen Elizabeth II Bridge, 2007.

A Christmas card sent to the staff by the management in 1987.

The longest-serving member of staff, Vic Hicks, started work at the Dartford Tunnel as a traffic-control man in February 1964, and retired as a senior traffic officer in April 2005 after serving forty-one years and three months.

The Tale of Don Stevens

There's been some tales at DRC but this one is the best
It's about the day that Dr Death disappeared into the west.
He wasn't following the setting sun or blazing the Oregon Trail
Nor like that man Columbus with a full ship under sail.
Nor like Captain Ahab on the Pequod (that's a whaler)
Don disappeared into the west on the back of someone's trailer.

He was walking across the plaza on his way back from break
But he had his eyes down to the ground, that was his big mistake.
He had stopped outside a toll booth to let a vehicle pass
But he did not see the trailer that tipped him on his arse.
It happened very quickly it didn't take a minute
The van with trailer drove away with the doctor lying in it.
His arms and legs were waving and no one heard him shout
But Alan Wickham saw him and so the call went out.

Rover at the garage into the west now go
There's a trailer with the doctor in and the driver doesn't know.
Don was being bounced around arms and legs were flailing
And a Rover was in hot pursuit with blues and two wailing.
The van stopped at Essex the driver got a fright
To see the doctor lying there just imagine what a sight.

Don was taken to the hospital the nurses there cried bingo
We can save an X-ray just hold him to the window.
They bandaged up his broken bone and sent him on his way
It will be a while before we see him next, he will be sick now until May.
And don't wonder why the van's rear mirror was not used to good effect
It was as everybody knows that vampires don't reflect.

                                            Peter Lewis

Other titles published by The History Press

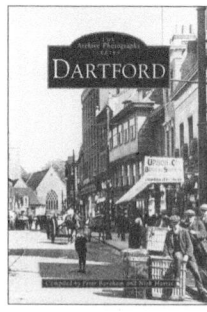

Dartford
PETER BOREHAM AND NICK HARRIS

This book is part of the Archive Photographs series, which uses old photographs and archived images to show the history of various local areas in Great Britain, through their streets, shops, pubs and people. This book is guaranteed to fascinate anyone who has lived in Dartford or its environs, and also people new to the area.

978 0 7524 0141 6

Rainham and Wennington Memories
CECILIA PYKE

The villages of Rainham and Wennington are closely linked, with the area having been shaped into the place it is today by the people, some of whose experiences and reminiscences are recorded here. This book brings together the personal memories of people who have lived and worked in Rainham and Wennington, vividly recalling childhood and working life, shops and entertainment, and the war years. The many absorbing stories are complemented by around 100 photographs.

978 0 7524 3671 5

Gravesend: Then & Now
EVE STREETER

This book is part of the Then & Now series, which matches old photographs with modern ones taken from the same camera locations to demonstrate the changes that have occurred over the years. The images include streets, buildings, people and social activities in each local area. This will be a revealing and intriguing read for locals and historians alike.

978 0 7524 2236 7

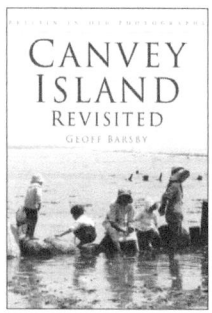

Canvey Island Revisited
GEOFF BARSBY

Canvey Island's location has contributed to its intriguing past. All over there is evidence of its history, from the pillboxes built during the Second World War to fairground attractions that show its prominence as a Victorian resort. Following on from the first two collections, this new volume revisits the Canvey Island of yesteryear with 200 fascinating photographs and postcards. As well as demonstrating the diversity of islanders throughout history, this edition pays particular attention to the North Sea floods of 1953.

978 0 7524 3984 6

Visit our website and discover thousands of other History Press books: **www.thehistorypress.co.uk**